I have a number of pets.

Here are just eight of them.

I like these eight pets.

I like this pet the best.
What pet do you think it is?
I will help you tell what it is!

My pet's skin is as green as grass.
Its two back legs are big.

Where is my pet's home?

Its home is a pond.

But it sleeps on my bed.

My pet can jump and swim.

It can go "ribbet, ribbet."

My pet has bugs for dinner.
Yuck!

Is my pet a sheep, a snake,
or a hog?

Is it a cat, a hen, or a dog?

No. My best pet is a frog!

Its name is Herman.

The End

Understanding the Story

Questions are to be read aloud by a teacher or parent.

1. What pet does the boy like best? (the frog)

2. When did you learn the answer to the question *What Is My Pet?* (Answers will vary.)

3. How does the boy help you figure out the answer? (Possible answer: by giving a lot of clues)

Saxon Publishers, Inc.
Editorial: Barbara Place, Julie Webster, Grey Allman, Elisha Mayer
Production: Angela Johnson, Carrie Brown, Cristi Henderson

Brown Publishing Network, Inc.
Editorial: Marie Brown, Gale Clifford, Maryann Dobeck
Art/Design: Trelawney Goodell, Camille Venti, Andrea Golden
Production: Joseph Hinckley

© Saxon Publishers, Inc., and Lorna Simmons

All rights reserved. No part of this publication may be reproduced, stored in a retrieval system, or transmitted in any form by any means, electronic, mechanical, photocopying, recording, or otherwise, without the prior written permission of the publisher. Address inquiries to Supervising Copy Editor, Saxon Publishers, Inc., 2450 John Saxon Blvd., Norman, OK 73071.

Printed in the United States of America
ISBN: 1-56577-962-2
Manufacturing Code: 01S0402
2 3 4 5 6 7 8 9 10 BSP 06 05 04 03 02

SAXON
Phonics and Spelling K

Phonetic Concepts Practiced

Review

ISBN 1-56577-962-2

Grade K, Decodable Reader 16
First used in Lesson 140

A Day at the Fair

written by Lisa Shulman
illustrated by Viki Woodworth

THIS BOOK IS THE PROPERTY OF:

STATE_____
PROVINCE_____
COUNTY_____
PARISH_____
SCHOOL DISTRICT_____
OTHER_____

Book No. _____
Enter information
in spaces
to the left as
instructed

ISSUED TO	Year Used	CONDITION	
		ISSUED	RETURNED

PUPILS to whom this textbook is issued must not write on any page or mark any part of it in any way, consumable textbooks excepted.

1. Teachers should see that the pupil's name is clearly written in ink in the spaces above in every book issued.
2. The following terms should be used in recording the condition of the book: New; Good; Fair; Poor; Bad.